Exploring Science

Focus on the NEXT GENERATION SCIENCE STANDARDS For States, By States

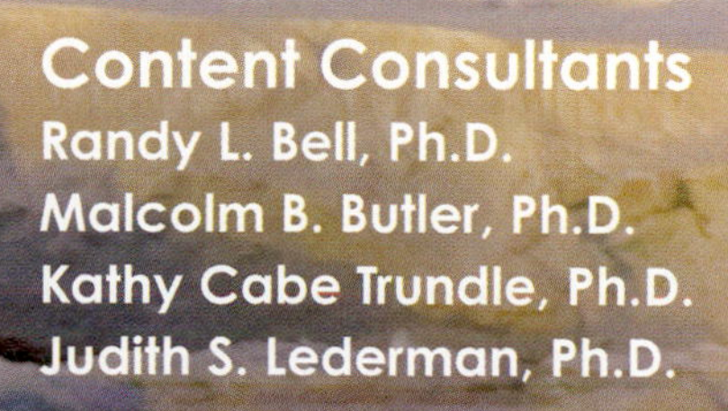

Content Consultants
Randy L. Bell, Ph.D.
Malcolm B. Butler, Ph.D.
Kathy Cabe Trundle, Ph.D.
Judith S. Lederman, Ph.D.

On the Cover
A steep mountain in Antarctica is almost covered with ice and snow. Chunks of ice float in the ocean water.

Physical Science

Life Science

Earth Science

Physical Science

Structure and Properties of Matter

Everything you see in this picture is matter.

Matter

Welcome to the American West! Look at the beautiful scenery. Did you know that everything in the picture is made of matter? **Matter** is anything that takes up space. You can describe matter. You can tell what makes one kind of matter different from another.

NEXT GENERATION SCIENCE STANDARDS | DISCIPLINARY CORE IDEAS
PS1.A: Structure and Properties of Matter
Different kinds of matter exist and many of them can be either solid or liquid, depending on temperature. (2-PS1-1)

The rock is a solid.
The water is a liquid.

Wrap It Up!

1. How would you describe the rock and the water in the picture?
2. Observe a pencil. Is the pencil more like the rock or the water? Explain.

My science notebook

Liquids

Water is a liquid when its temperature is above freezing. A **liquid** takes the shape of its container. It does not have a shape of its own.

What happens if a liquid doesn't have a container? It spreads out. A liquid does not have a definite shape when it's not in a container.

Water takes the shape of the river bed.

NEXT GENERATION SCIENCE STANDARDS | DISCIPLINARY CORE IDEAS
PS1.A: Structure and Properties of Matter
Different kinds of matter exist and many of them can be either solid or liquid, depending on temperature. (2-PS1-1)

This water takes the shape of the trough that contains it.

Wrap It Up!

1. What is one property of liquids?
2. Honey is in a bottle shaped like a bear. What shape does the honey have?

My science notebook

Solids

When the temperature is below freezing, water freezes. It becomes solid. A **solid** is matter that has its own shape. It also takes up space.

NEXT GENERATION SCIENCE STANDARDS | DISCIPLINARY CORE IDEAS
PS1.A: Structure and Properties of Matter
Different kinds of matter exist and many of them can be either solid or liquid, depending on temperature. (2-PS1-1)

Ice is solid water.

Wrap It Up!

1. What is a solid?
2. How are solids and liquids different from each other?

My science notebook

Investigate

Solids and Liquids

How is a liquid different from a solid?

Shape is one property of solids and liquids. In this investigation, you will observe the shape of a liquid. You will also observe the shape of a solid.

Materials

two plastic cups

cylinder of water

marble

NEXT GENERATION SCIENCE STANDARDS | DISCIPLINARY CORE IDEAS
PS1.A: Structure and Properties of Matter
Matter can be described and classified by its observable properties. (2-PS1-1)

My science notebook

1. Observe the shape of the water in the cylinder. Record your observations.

2. Pour the water from the cylinder into one of the cups. Record your observations.

3. Observe the shape of the marble. Record your observations.

4. Put the marble into the cylinder and then into the empty cup. Observe the marble's shape each time. Record your observations.

Wrap It Up!

1. What kind of matter is water? How do you know?
2. What kind of matter is a marble? How do you know?

My science notebook

Properties

You can describe the objects you see. When you describe something, you talk about its properties. A **property** is something about an object you can observe with your senses.

Shape is one property of matter. These bolo ties have different shapes.

rectangle

oval

NEXT GENERATION SCIENCE STANDARDS | DISCIPLINARY CORE IDEAS
PS1.A: Structure and Properties of Matter
Matter can be described and classified by its observable properties. (2-PS1-1)

Wrap It Up!

1. What is a property?
2. Why is shape a property?

My science notebook

Color

Color is another property of matter. You can use color to tell one object from another.

NEXT GENERATION SCIENCE STANDARDS | DISCIPLINARY CORE IDEAS
PS1.A: Structure and Properties of Matter
Matter can be described and classified by its observable properties. (2-PS1-1)

Wrap It Up!

1. How can you describe one boot to tell it from the others?
2. Which of your senses do you use to describe color?

My science notebook

Texture

Look at all the objects in the raft. What do these different objects feel like? **Texture** is a property. It is the way an object feels. Touch the paddle, and it feels smooth. It has a smooth texture. Touch the rope, and it feels rough. It has a rough texture.

Look at the picture. What else has a rough or smooth texture?

NEXT GENERATION SCIENCE STANDARDS | DISCIPLINARY CORE IDEAS
PS1.A: Structure and Properties of Matter
Matter can be described and classified by its observable properties. (2-PS1-1)

SCIENCE in a SNAP

Time to Sort

1 Find objects like these in your classroom.

2 Sort the objects by texture. Are they rough or smooth?

? **Tell what other objects in your classroom have a rough or smooth texture.**

Wrap It Up!

1. What is texture?
2. Choose an object in your classroom. Describe its texture.

My science notebook

Hard and Soft

You can tell whether objects feel hard or soft. Some things have both hard parts and soft parts.

Look at the picture. The horse's hair feels soft. The horse has hard hooves.

Can you find other objects that are hard or soft in the picture?

NEXT GENERATION SCIENCE STANDARDS | DISCIPLINARY CORE IDEA
PS1.A: Structure and Properties of Matter
Matter can be described and classified by its observable properties. (2-PS1-1)

SCIENCE in a SNAP

Sort Some More

1 Find objects like these at home.

2 Sort the objects by texture. Are they hard or soft?

? **Tell what other objects in your home are hard or soft or both.**

Wrap It Up!

1. Describe the properties of objects that are hard.
2. In your home, what objects do you want to be soft? Why?

My science notebook

Bend and Stretch

Look at the rope in this picture. It is flexible. **Flexible** materials can bend without breaking. The rope bends easily. These climbers can loop it through hooks in the rock as they climb.

Some materials can stretch, too. The climbers pull the rope to stretch it tight.

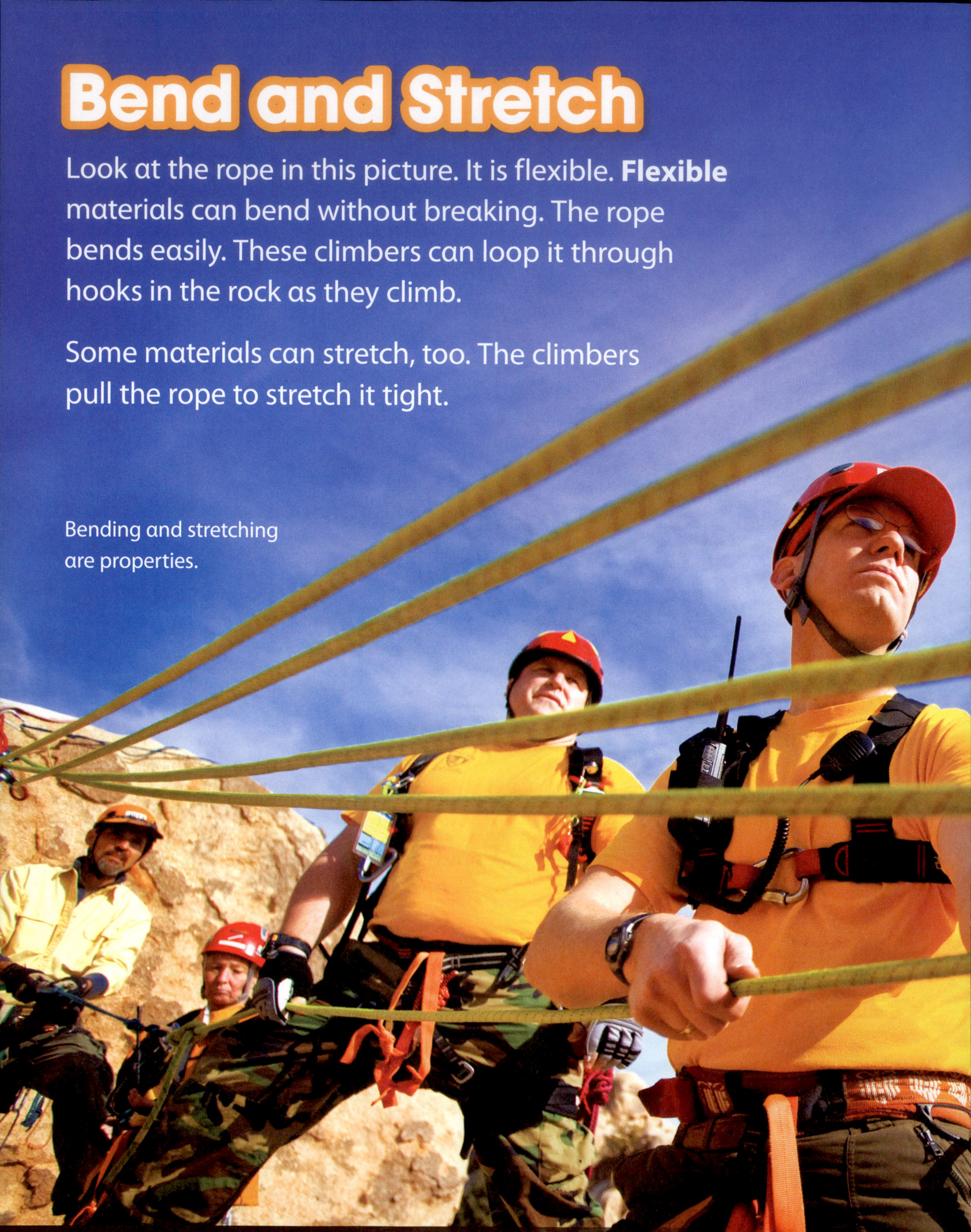

Bending and stretching are properties.

NEXT GENERATION SCIENCE STANDARDS | DISCIPLINARY CORE IDEAS
PS1.A: Structure and Properties of Matter
Matter can be described and classified by its observable properties. (2-PS1-1)

The cowboy's rope can bend into a loop.

SCIENCE in a SNAP

Bend and Stretch

1 Rubber bands are very flexible. They bend and stretch. Stretch and bend a rubber band to see what a flexible object feels like.

2 Find other objects in your classroom that are flexible. Which objects bend? Which objects stretch?

Tell what objects in your home are flexible.

Wrap It Up!

1. What does it mean to be flexible?
2. Is paper flexible? Explain why or why not.

My science notebook

Sink and Float

Objects can sink or float in liquids. Whether something sinks or floats is a property. If you throw a heavy rock into a pond, it will sink to the bottom of the pond. Other objects may float. The boat floats on top of the water.

The anchor sinks to the bottom of the water.

NEXT GENERATION SCIENCE STANDARDS | DISCIPLINARY CORE IDEAS
PS1.A: Structure and Properties of Matter
Different properties are suited to different purposes. (2-PS1-2), (2-PS1-3)

The boat floats because of the material it is made of and its shape.

Wrap It Up!

1. Explain what sink and float mean.
2. List two things that float.
3. List two things that sink.

Plan and Investigate

You have observed properties of matter. When you classify objects, you sort them into groups. In this investigation, you will sort objects. You will classify the objects by their properties.

1. **Ask a question.**
   How can you classify materials by their observable properties?

2. **Plan and carry out an investigation.**

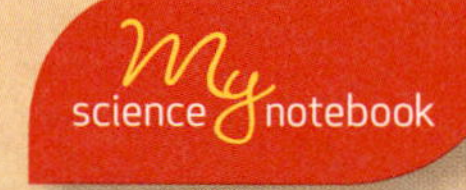

   Make a plan. Write down your plan. Gather materials.

   Now carry out your plan. Record your data in a chart.

NEXT GENERATION SCIENCE STANDARDS | PERFORMANCE EXPECTATION
2-PS1-1. Plan and conduct an investigation to describe and classify different kinds of materials by their observable properties.

3 Analyze and interpret data.

Look at your data. What properties did you observe? How did you classify objects by those properties?

4 Share and explain your results.

Tell others how your investigation worked. Explain how your results answer the question.

Investigate

Materials That Absorb

Which materials absorb the most water?

If you spill water on a table, the water sits on the surface. If you wipe up the spill with a towel, the water soaks into the towel. The towel **absorbs** the water. In this investigation, you will compare how materials absorb water.

Materials

NEXT GENERATION SCIENCE STANDARDS | PERFORMANCE EXPECTATION
2-PS1-2. Analyze data obtained from testing different materials to determine which materials have the properties that are best suited for an intended purpose.

1. Use the measuring cup to pour 50 mL into the cup labeled **paper.**

My science notebook

2. Observe the paper. What do you think will happen when you dip a corner of the piece of paper into the water for 30 seconds? Record your prediction.

3. Dip a corner of the paper into the water for 30 seconds. Use a timer to measure the time. Then take the paper out of the cup.

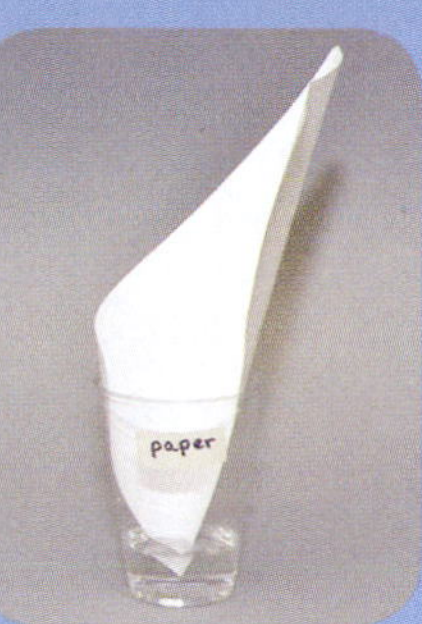

4. Repeat steps 1–3 with each labeled cup and each of the other materials. Then put the cups in order from least amount of water to most amount of water. Record your observations.

Wrap It Up!

1. Which cup had the least water left in it? Which cup had the most water left in it?
2. Which material would work best for cleaning up spilled liquid? Explain.
3. Did your results support your predictions? Explain.

My science notebook

Build It

This huge arch is the entrance to Yellowstone National Park. Look closely. It is made up of many smaller pieces.

A small set of pieces can be used to build many different objects. Many smaller stone pieces were used to make this big arch.

NEXT GENERATION SCIENCE STANDARDS | DISCIPLINARY CORE IDEAS
PS1.A: Structure and Properties of Matter
A great variety of objects can be built up from a small set of pieces. (2-PS1-3)

Other objects are made using many stone pieces, such as this lodge.

Wrap It Up!

1. Can you think of other objects people make from pieces of stone?
2. What kinds of toys can you use to build many different shapes from just a few types of pieces?

My science notebook

Make Observations

Look closely at the objects all around you. You can see that the objects are made of smaller parts. Some of those smaller parts might be alike. You can find similar parts used to build different, bigger things.

1 Ask a question.

How can the same materials be used to make different objects?

2 Carry out an investigation.

My science notebook

Gather materials. Use the materials to build an object. Observe what you have made.

Record your observations in your science notebook.

3 Observe and record.

Trade objects with a partner. Observe what he or she has built. Record your observations.

Take the object apart. Build something new with the parts. Observe what you have built. Record your observations.

4 Analyze and interpret data.

Look at your observations. How did you and your partner use the same materials differently?

NEXT GENERATION SCIENCE STANDARDS | PERFORMANCE EXPECTATION
2-PS1-3. Make observations to construct an evidence-based account of how an object made of a small set of pieces can be disassembled and made into a new object.

5 Use evidence.

How do your results answer the question?

6 Share and explain your results.

Tell others how your investigation worked. Give evidence to explain how your observations help you answer the question.

How could you use the same type of pieces to build a fence around the cabin?

Cooling

The mountain lion cubs in the picture are walking on ice. Cold temperatures cause liquid water to freeze. When water freezes, it turns into ice. It changes from a liquid to a solid.

NEXT GENERATION SCIENCE STANDARDS | DISCIPLINARY CORE IDEAS
PS1.A: Structure and Properties of Matter
Different kinds of matter exist and many of them can be either solid or liquid, depending on temperature. (2-PS1-1)
PS1.B: Chemical Reactions
Heating or cooling a substance may cause changes that can be observed. (2-PS1-4)

SCIENCE in a SNAP

Cooling Water

1 Observe the shape of water in a cup. Use a piece of clay to make a container. Pour water into the clay container. Put the container into a freezer for 4 hours. Predict what will happen to the water.

2 Take the ice out of the container and put it in a cup. Observe the shape of the ice in the cup.

? **Did your results support your prediction? How is the ice different from the liquid water?**

Wrap It Up!

1. How does water change when it is cooled to freezing?
2. Describe what a cup of water would look like if it were halfway frozen.

My science notebook

Heating

Warm temperatures can also cause matter to change. When water freezes into ice, heat can cause it to melt into a liquid again. When ice melts into water, cooling can freeze it into solid ice again. The change can happen over and over.

Warming will change this frozen snow to liquid water droplets.

NEXT GENERATION SCIENCE STANDARDS | DISCIPLINARY CORE IDEAS
PS1.B: Chemical Reactions
Heating or cooling a substance may cause changes that can be observed. Sometimes these changes are reversible, and sometimes they are not. (2-PS1-4)

Heating Ice

1 Make a small container out of foil. Place an ice cube in the container.

2 Predict what will happen to the ice cube. Record your prediction. After 1 hour, observe the ice cube.

? **Do your results support your prediction? How has the shape of the ice cube changed?**

Wrap It Up!

1. Explain how ice changes when it is heated.
2. What would happen if you put the foil container at the end of your Science in a Snap investigation into the freezer?

My science notebook

Change It?

While cooking on a campfire, you can learn about matter. Cooking changes matter. Some matter changes but cannot change back the way water does. Your breakfast eggs start out as a liquid. As they cook, they become a solid. The change cannot be reversed.

NEXT GENERATION SCIENCE STANDARDS | DISCIPLINARY CORE IDEAS
PS1.B: Chemical Reactions
Heating or cooling a substance may cause changes that can be observed. Sometimes these changes are reversible, and sometimes they are not. (2-PS1-4)

BEFORE Raw eggs are liquid. They take the shape of their container.

AFTER Heat cooks the eggs. Cooked eggs become solid and have their own shape. Cooked eggs cannot change back into a liquid.

Wrap It Up!

1. Describe the change that cannot be reversed between a raw egg and a cooked egg.
2. What caused the changes?

Make an Argument

Heating and cooling materials can change them. The pictures on these pages show something that has changed because of heating or cooling.

ice

NEXT GENERATION SCIENCE STANDARDS | PERFORMANCE EXPECTATION
2-PS1-4. Construct an argument with evidence that some changes caused by heating or cooling can be reversed and some cannot.

bread

popcorn

Wrap It Up!

1. Describe the change in each of the three pictures that was caused by heating or cooling.
2. Which changes can be reversed? Which changes cannot be reversed?
3. Give evidence to support your answer for each picture.

My science notebook

Science Career

Materials Scientist

What would you use to make a bucket? You would need to choose material that does not leak. You would have to use something that does not absorb liquid.

Dr. Ainissa Ramirez is a materials scientist. A materials scientist figures out what materials to use to build things. Dr. Ramirez studies properties. She develops and tests new materials. She also finds fun ways to show people amazing things about different materials.

NEXT GENERATION SCIENCE STANDARDS | CONNECTIONS TO NATURE OF SCIENCE
Science Models, Laws, Mechanisms, and Theories Explain Natural Phenomena.
Science searches for cause and effect relationships to explain natural events.

Dr. Ainissa Ramirez makes fun videos for young people. She shows neat examples of the properties of materials!

What makes this metal strip curl? Heat! And when it cools, it will straighten out again!

Life Science

Interdependent Relationships in Ecosystems

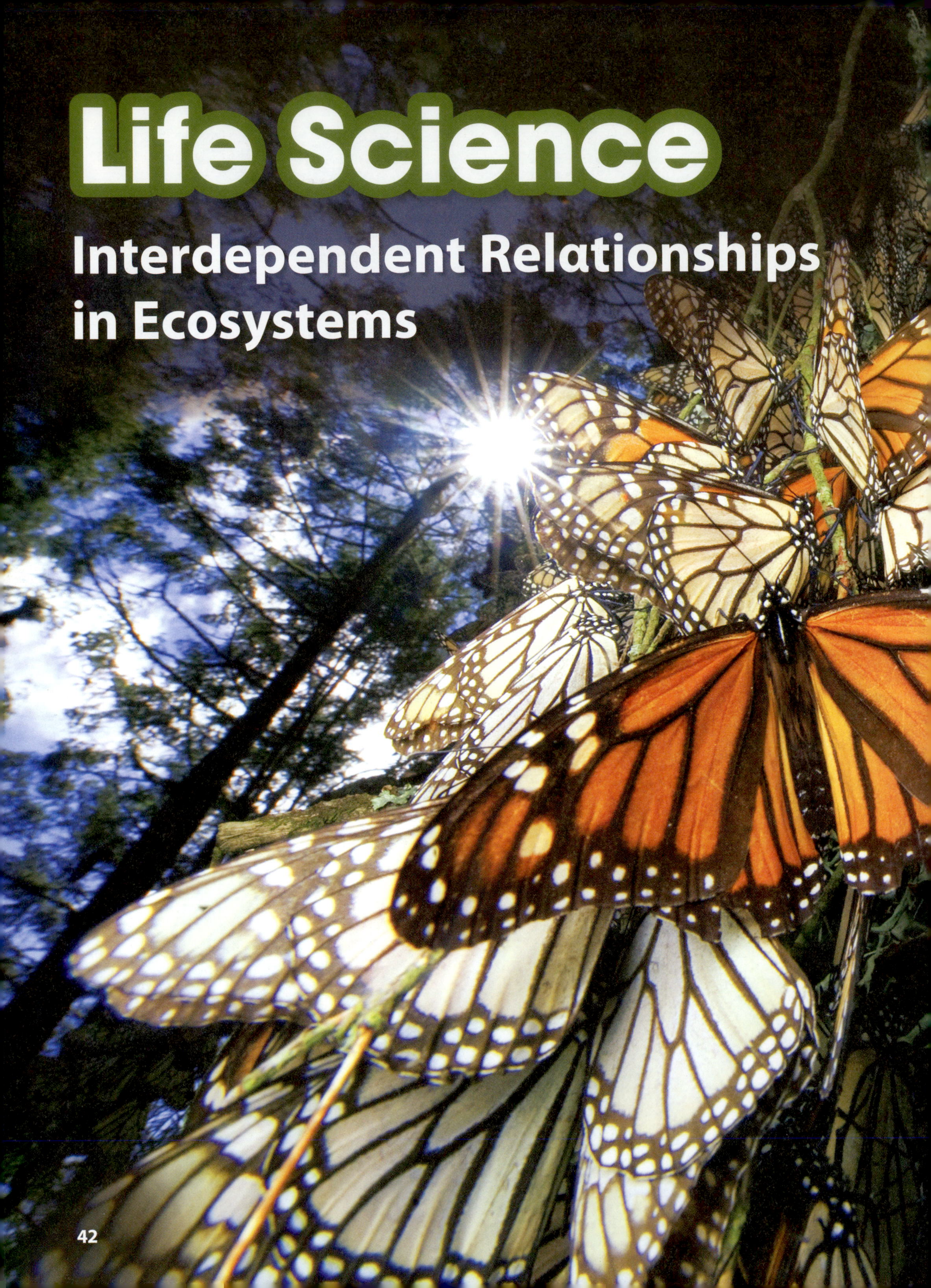

Monarch butterflies rest on a tree in Mexico.

What Plants Need

Plants need light and water. Plants **depend** on light and water to live and grow.

Some plants need a lot of light. These poppies growing in a field get sunlight all day.

Poppies grow well in direct sunlight.

NEXT GENERATION SCIENCE STANDARDS | DISCIPLINARY CORE IDEAS
LS2.A: Interdependent Relationships in Ecosystems
Plants depend on water and light to grow. (2-LS2-1)

The poppies also need rain every few days to survive and grow.

Rain gives the foxglove plant the water it needs to live.

Wrap It Up!

1. What are some things that plants depend on to live and grow?
2. Many people grow plants indoors. How can plants grow inside where it does not rain?

My science notebook

Investigate

Plants and Light

? What happens if radish plants do not get enough light?

You know that plants need light to grow. Now you will investigate how plants grow with different amounts of light.

Materials

radish plants

tape

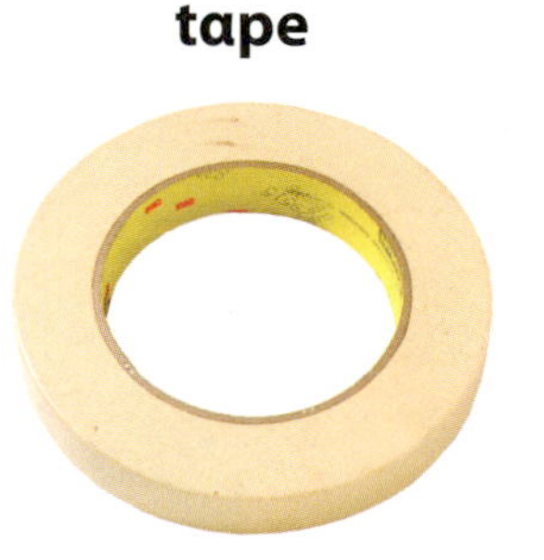

water

spoon

My science notebook

1 Label the cups *Sunlight* and *No Sunlight*. Observe the plants in both cups. Record your observations.

2 Place the *Sunlight* cup in a sunny place. Predict what will happen. Record your prediction in your science notebook.

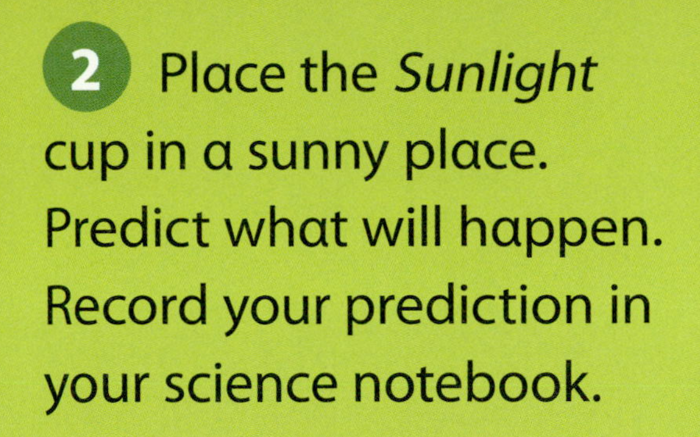

3 Place the *No Sunlight* cup in a dark place. Predict what will happen. Record your prediction in your science notebook.

4 Give both plants two spoonfuls of water every day. Observe the plants every day. Record what you observe.

Wrap It Up!

1. What did you keep the same for both plants?
2. What did you change between the two plants?
3. How did your predictions compare to your observations?

My science notebook

Plan and Investigate

You have investigated whether plants need light to grow. Now you will make a plan and investigate what happens if plants do not get water.

1 Plan an investigation.

Think about how to do your investigation. What materials will you need? How many plants should you test? How will you know that water makes the difference in how your plants grow?

Write down your plan. Draw a picture of what your investigation will look like. Label your drawing.

2 Conduct an investigation.

Carry out your investigation. Record your data in your science notebook.

NEXT GENERATION SCIENCE STANDARDS | PERFORMANCE EXPECTATION
2-LS2-1. Plan and conduct an investigation to determine if plants need sunlight and water to grow.

3. **Review your results.**
Look at your data. Did one plant grow better than the others? Do your results answer the question?

4. **Share your results.**
Tell others how your investigation worked. Use evidence to explain how your results answer the question.

Rain provides this fuchsia plant with the water it needs to survive and grow.

Animals Pollinate Flowers

Flowering plants make **pollen.** When pollen is moved from one plant to another, seeds can form. Seeds can grow into new plants.

Some animals get food from flowers. Pollen sticks to an animal when it visits a flower for food. The animal **pollinates** the next flower it visits. Some pollen from the first flower is left on the second flower. Some plants depend on animals for pollination.

Hummingbirds pollinate flowers as they collect food.

NEXT GENERATION SCIENCE STANDARDS | DISCIPLINARY CORE IDEAS
LS2.A: Interdependent Relationships in Ecosystems
Plants depend on animals for pollination or to move their seeds around. (2-LS2-2)

Pollen sticks to the moth's feet.

Ants carry pollen from flower to flower. You can see the yellow pollen sticking to the ant's back.

Wrap It Up!

1. Why is pollination important to plants?
2. What are some ways that animals move pollen from flower to flower?

My science notebook

Save the Bees!

Do you like strawberries? Watermelon? Thanks to bees, we have fruits and other foods to eat. Bees pollinate many plants that make food. Some pollinated flowers transform into fruits that people eat. Dino Martins is a scientist who studies bees and other insects.

Problem

Dino Martins is worried. There are fewer bees today than there were in the past. Fewer bees mean fewer fruits and other foods. People need food, so people need bees, too.

This Braunsapis bee is pollinating coriander flowers. People grind and use coriander seeds as a spice.

NEXT GENERATION SCIENCE STANDARDS | DISCIPLINARY CORE IDEAS
LS2.A: Interdependent Relationships in Ecosystems
Plants depend on animals for pollination or to move their seeds around. (2-LS2-2)

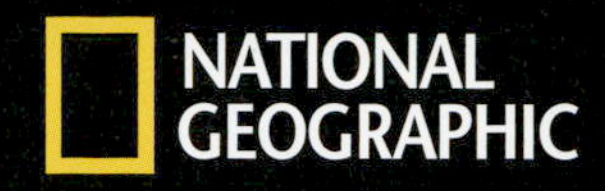

Explorer

Dino Martins says, "Spend five minutes a day with an insect. It will change your life." His work with insects is changing many lives.

There are many kinds of bees. Dino Martins studies bees in Africa.

Amegilla bee

Stingless bee and Nomia bee

Carpenter bee

Solution

Dino Martins helps people protect bees. That way, there will be more bees to pollinate flowers. He teaches farmers that bees pollinate crops. Farmers who protect bees harvest more food. We need plenty of food for people to eat.

Some bees' habitats have been destroyed. Dino Martins shows people how to make bee houses. Bee houses help bees survive.

A bee house on a farm gives bees shelter near crops that they pollinate.

Dino teaches people about the types of bees that help the plants on their farms.

On a field trip Dino helps students collect insects so that they can look at them up close.

Wrap It Up!

1. Why are bees important?
2. What might happen if there were no bees?
3. Why do you think Dino Martins' work is important?

My science notebook

Animals Spread Seeds

Many plants make seeds that grow new plants. Seeds need their own space to grow. A seed may not be able to grow too close to its parent plant. It may not get enough light or water there.

How can a seed move to a new place? Animals carry some seeds to new places. Many plants depend on animals to move their seeds.

This red squirrel is burying a nut to eat later. If the squirrel forgets the nut, a new tree may grow.

These berries have seeds inside. The waxwing will eat a berry. The waxwing will drop the seed in a new place.

NEXT GENERATION SCIENCE STANDARDS | DISCIPLINARY CORE IDEAS
LS2.A: Interdependent Relationships in Ecosystems
Plants depend on animals for pollination or to move their seeds around. (2-LS2-2)

Seeds with hooks and barbs are called burrs. Burrs stick to fur. The cow will carry the seeds to a new place.

Wrap It Up!

1. What are some ways seeds move to new places?
2. How do you think the burrs got onto the cow's head?

My science notebook

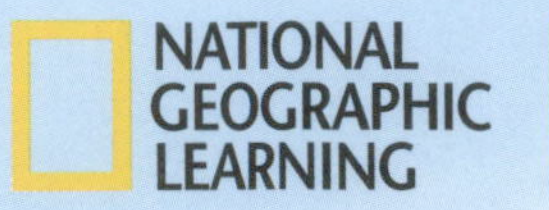

Think Like a Scientist

Develop a Model

It is close to the end of summer. The burdock plant in the picture has produced a special covering for its seeds. The covering is called a burr. Look closely. The burr has long spiky thorns with hooks on the end. Many of the seeds from the parent plant have traveled quite far away to produce new plants. How did the seeds move?

This is what the burrs look like in fall.

NEXT GENERATION SCIENCE STANDARDS | DISCIPLINARY CORE IDEAS
ETS1.B. Developing Possible Solutions
Designs can be conveyed through sketches, drawings, or physical models. These representations are useful in communicating ideas for a problem's solutions to other people. (secondary to 2-LS2-2)

NEXT GENERATION SCIENCE STANDARDS | PERFORMANCE EXPECTATION
2-LS2-2. Develop a simple model that mimics the function of an animal in dispersing seeds or pollinating plants.

Develop a model.

When scientists want to study how something in real life works, they sometimes make a model. You can, too. Look at the burdock seeds again. Then draw or write out a model of how burdock seeds could travel far from their parent plants.

Conduct an investigation.

Now you can conduct an investigation to see if your model is on the right track.

1. Stick the hook tape to the feather. Pinch them together between your fingers.
2. Carry the feather to a new place. Shake the feather. How is this like what a bird might do? Observe and record what happens.

3. Pick the tape off with your fingers. How is this like what a bird might do? Observe and record what happens.

4. Repeat steps 2 and 3 with the fake fur and the leather.

Explain your model.

Based on the data you collected, how well did your model match your observations? Do you need to revise your model? Make changes if you need to. Then use your model to explain how sticky seeds might be moved from one place to another.

Living Things Are Everywhere

Many kinds of living things live all over planet Earth. They live on land and in water. Forests, deserts, oceans, and ponds are full of living things. Different kinds of plants and animals live in different places.

The fish in this kelp forest live in water all of the time.

NEXT GENERATION SCIENCE STANDARDS | DISCIPLINARY CORE IDEAS
LS4.D: Biodiversity and Humans
There are many different kinds of living things in any area, and they exist in different places on land and in water. (2-LS4-1)

The hippopotamus spends time on the land and in the water.

The chameleon lives on land all of the time.

Wrap It Up!

1. Where on Earth do plants and animals live?
2. Describe how the animals and habitats shown are alike and different.

My science notebook

Living Things on the Coast

Many plants and animals live on the sandy coast. A **coast** is land that is next to the ocean. You can see sea grass, a crab, a turtle, and an egret. These plants and animals depend on the sandy coast for survival.

The ghost crab eats clams and newly hatched turtles. It even eats other crabs. The egret eats snakes, insects, and fish. Insects and other small animals hide in the sea grass.

The ghost crab can burrow into the sand to remain hidden from other animals, such as the egret.

This young sea turtle has just hatched. It crawls quickly to the ocean. The turtle is safer in the ocean. It can find food and shelter there.

NEXT GENERATION SCIENCE STANDARDS | DISCIPLINARY CORE IDEAS
LS4.D: Biodiversity and Humans
There are many different kinds of living things in any area, and they exist in different places on land and in water. (2-LS4-1)

The grass provides food and shelter for some animals.

The egret will use its sharp bill to catch other animals.

Wrap It Up!

1. Choose an animal that lives on the coast. Tell how it depends on another animal or a plant.
2. What might happen to some animals if the sea grass died?
3. Why do you think the young turtle crawls quickly to the ocean?

My science notebook

Living Things in a Wetland

A **wetland** is land that is covered with water some of the time. Plants and animals here are suited to life in this wet place. The heron has long legs. It wades in the water as it hunts for fish.

The alligator moves easily through the water to hunt. It may catch fish and turtles with its strong jaws. It may also eat the heron if it can catch it.

Vultures eat animals that have died. Vultures can find plenty of food in this wetland.

These vultures eat whatever they can find. They might eat the alligator's leftovers.

NEXT GENERATION SCIENCE STANDARDS | DISCIPLINARY CORE IDEAS
LS4.D: Biodiversity and Humans
There are many different kinds of living things in any area, and they exist in different places on land and in water. (2-LS4-1)

Wrap It Up!

1. What is a wetland?
2. What can you observe from the picture that makes the heron, alligator, and vulture suited to life in a wetland?

My science notebook

The great blue heron can fly away if the alligator gets too close.

The alligator hunts mostly in the water. It gets warm by lying on land in the sunlight.

Living Things in a Grassland

Grasslands are found all over the world. The grassland in this picture is in Australia. Compared to wetlands, grasslands are dry places. They get enough water for grasses to grow. They do not get enough water for many trees to grow.

Look at these pictures of some animals that live in Australian grasslands. Why do you think these animals are all similar in color? They all blend well with the grass. This coloring helps them hunt for food without being seen by other animals.

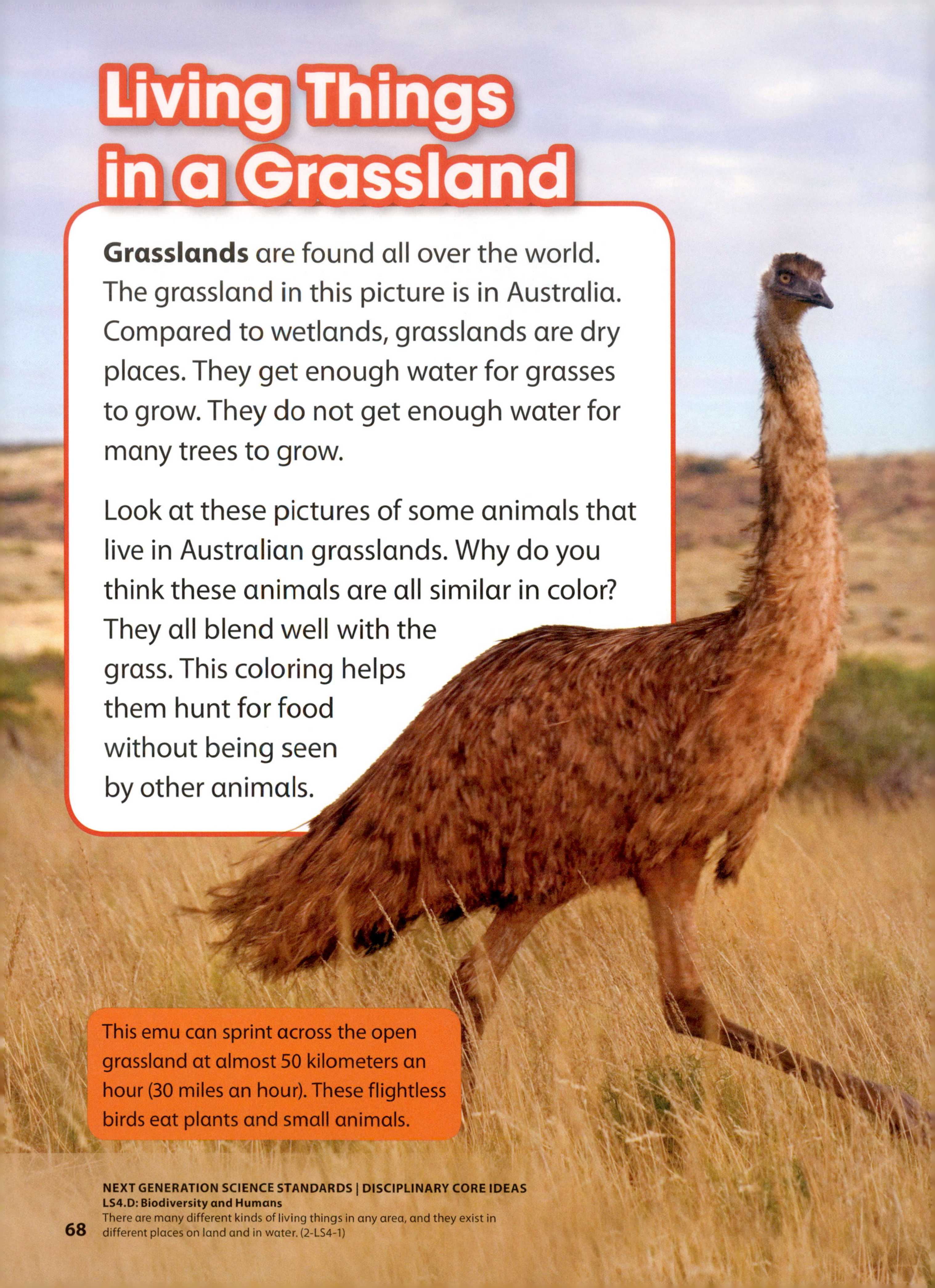

This emu can sprint across the open grassland at almost 50 kilometers an hour (30 miles an hour). These flightless birds eat plants and small animals.

NEXT GENERATION SCIENCE STANDARDS | DISCIPLINARY CORE IDEAS
LS4.D: Biodiversity and Humans
There are many different kinds of living things in any area, and they exist in different places on land and in water. (2-LS4-1)

The dingo is a wild dog. Groups of dingoes hunt other animals that live in the grassland.

A red kangaroo eats mostly grasses and other plants. It can cover 8 meters (25 feet) in a single hop!

Wrap It Up!

1. How are grasslands different from wetlands?
2. What characteristics do the pictured animals have that help them live in a grassland?

Make Observations

Africa has different habitats. Some are hot and dry. Others are wet. Look at the map to find deserts, rain forests, and other habitats in Africa. Each habitat has many plants and animals that live there.

Look at the pictures of plants and animals from different habitats in Africa. Then answer the questions.

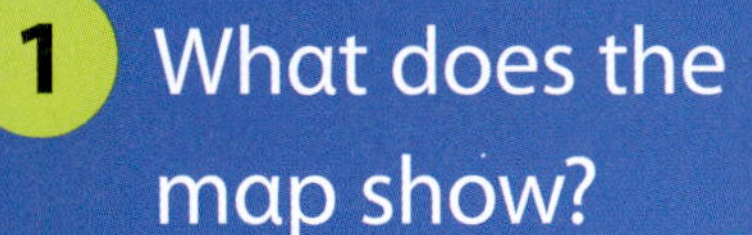

1. What does the map show?

2. Choose two habitats. Compare the plants and animals that live there. How are they alike? How are they different?

NEXT GENERATION SCIENCE STANDARDS | PERFORMANCE EXPECTATION
2-LS4-1. Make observations of plants and animals to compare the diversity of life in different habitats.

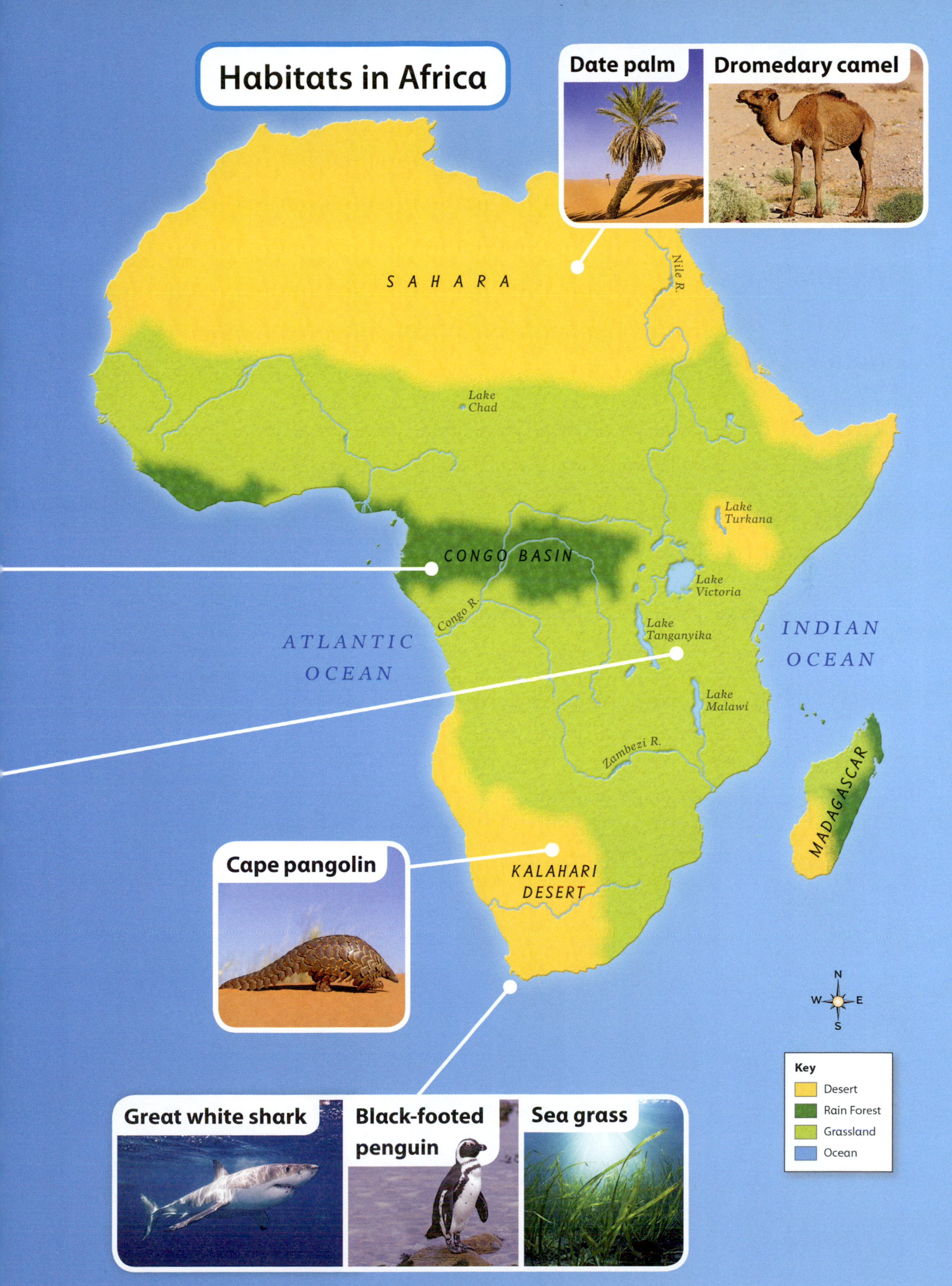
Habitats in Africa
Date palm
Dromedary camel
SAHARA
Nile R.
Lake Chad
Lake Turkana
CONGO BASIN
Lake Victoria
Congo R.
Lake Tanganyika
ATLANTIC OCEAN
INDIAN OCEAN
Lake Malawi
Zambezi R.
MADAGASCAR
Cape pangolin
KALAHARI DESERT
N
W
E
S
Key
Desert
Rain Forest
Grassland
Ocean
Great white shark
Black-footed penguin
Sea grass

Science Career

Field Biologist

A biologist is a scientist who studies living things. A field biologist studies plants or animals where they live.

Tim Laman is a field biologist and a photographer. He studies living things in places such as rain forests and coral reefs. He takes photos of the living things that he studies.

Tim does much of his research in places where the animals' homes are in danger. He hopes his research will make people want to take better care of natural places.

Tim Laman takes photos of birds from treetops. Here he is in New Guinea.

NEXT GENERATION SCIENCE STANDARDS | CONNECTIONS TO NATURE OF SCIENCE

Scientific Knowledge Is Based on Empirical Evidence

Scientists look for patterns and order when making observations about the world.

Explorer

Tim Laman loved to explore the mountains and oceans of Japan as a boy. He hopes the photos he takes help tell the stories of some of Earth's endangered species.

Blue bird of paradise

Greater bird of paradise

Wilson's bird of paradise

Earth Science

Earth's Systems: Processes that Shape the Earth

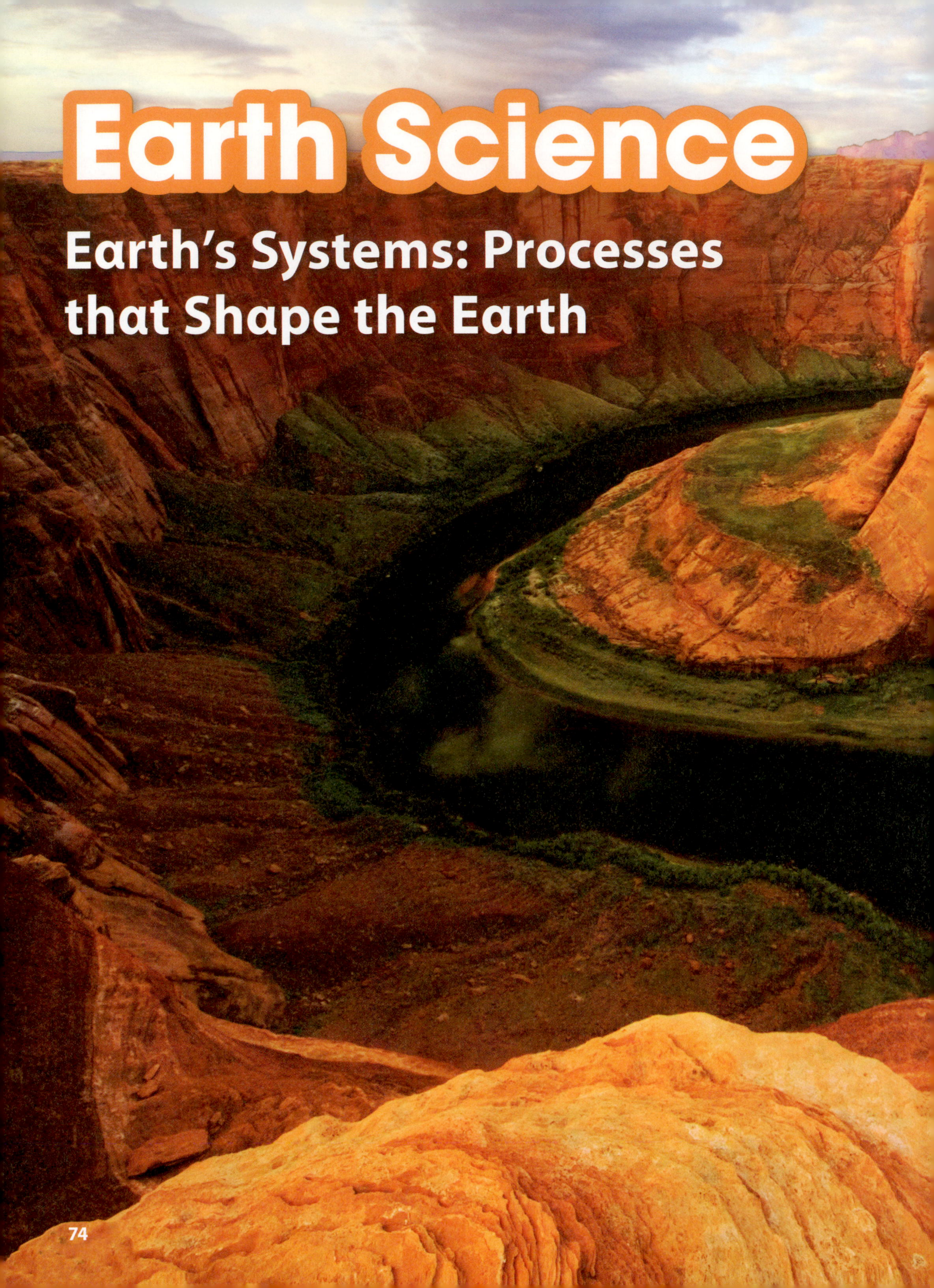

The Colorado River flows through the Grand Canyon in Arizona.

Earthquakes

Most people never feel the ground move beneath their feet. It can happen, though! During an **earthquake,** Earth's surface moves and shakes.

Earthquakes can change Earth's surface quickly. They can produce large cracks in the ground. Earthquakes can cause buildings, roads, and bridges to crumble. In 2010 an earthquake shook Haiti. Many buildings, such as the cathedral in the city of Port-au-Prince, were destroyed.

This picture shows what the Our Lady of Assumption Cathedral in Port-au-Prince looked like before the earthquake.

NEXT GENERATION SCIENCE STANDARDS | DISCIPLINARY CORE IDEAS
ESS1.C: The History of Planet Earth
Some events happen very quickly; others occur very slowly, over a time period much longer than one can observe. (2-ESS1-1)

Wrap It Up!

1. What happens during an earthquake?
2. Why might earthquakes be more dangerous to people in cities than to living things in the wild?

My science notebook

Volcanoes

There is melted rock called magma inside Earth. Magma, gases, and ash can **erupt** through a **volcano.** A volcano can change Earth's surface. The change can happen very fast.

Mount St. Helens was shaped like a cone. The volcano had erupted many times. Magma erupted up through the volcano. When magma reaches the surface it is called lava. The lava hardens into a layer of rock. Over time, the layers of rock built up to form the cone shape of the volcano.

In 1980 Mount St. Helens erupted again. This time the eruption was much bigger. A huge explosion flattened trees for miles around. A thick layer of ash covered the land. The explosion and ash left a lifeless landscape.

NEXT GENERATION SCIENCE STANDARDS | DISCIPLINARY CORE IDEAS
ESS1.C: The History of Planet Earth
Some events happen very quickly; others occur very slowly, over a time period much longer than one can observe. (2-EwSS1-1)

The 1980 eruption blasted away a large area at the top of the volcano. The shape of the volcano was changed very quickly.

Wrap It Up!

1. How can some eruptions cause rapid change to Earth's surface?
2. Melted rock from a volcano covers the land. It hardens into new rock. If a volcano erupts many times, how will the surface change over time?

My science notebook

Weathering and Erosion

Look at the unusual rock shapes in the picture. Processes called weathering and erosion carved them.

Weathering and erosion change Earth's surface. **Weathering** can break rock into smaller pieces. What happens to these pieces of rock? Wind and water can carry the pieces to new places. The moving of rocks, sand, or soil to a new place is called **erosion.**

NEXT GENERATION SCIENCE STANDARDS | DISCIPLINARY CORE IDEAS

ESS1.C: The History of Planet Earth
Some events happen very quickly; others occur very slowly, over a time period much longer than one can observe. (2-ESS1-1)

ESS2.A: Earth Materials and Systems
Wind and water can change the shape of the land. (2-ESS2-1)

This rock formation in England is called Durdle Door. The arch formed when parts of the rock were weathered and eroded away.

SCIENCE in a SNAP

Rocks Change

1 Put one piece of sandstone in a jar of water. Screw the lid on tight. Use a hand lens to observe the sandstone at the bottom of the jar.

2 Shake the jar hard. Observe with the hand lens again.

? **What changes do you observe? Explain how this investigation models weathering and erosion of rocks.**

Wrap It Up!

1. What are some causes of erosion?
2. Why does weathering need to occur before erosion can occur?

My science notebook

Wind Changes Land

When you feel wind on your face, you probably do not think it could change the land. Over a long time, though, wind does just that. Wind can wear away rock and change Earth's surface.

The process of weathering breaks apart or changes rocks. When rock pieces become tiny enough, the wind can carry them away. Over thousands of years, the shapes of the rocks can change.

Tiny pieces of rock are called **sediment.** Wind blows sediment against larger rocks. It wears them away.

NEXT GENERATION SCIENCE STANDARDS | DISCIPLINARY CORE IDEAS

ESS1.C: The History of Planet Earth
Some events happen very quickly; others occur very slowly, over a time period much longer than one can observe. (2-ESS1-1)

ESS2.A: Earth Materials and Systems
Wind and water can change the shape of the land. (2-ESS2-1)

Wind has weathered and eroded this rock so it looks smooth.

Weathering can wear away the sides of large rocks leaving tall shapes. Wind through the Grand Canyon left this sandstone shape.

Wrap It Up!

1. Do you think that wind changes the land quickly or slowly? Explain your answer.
2. What do you think will happen to the shape of the rock in the picture over a long period of time?

My science notebook

Water Changes Land

Moving water can also change Earth's surface. Just like wind, moving water picks up small rocks. The rocks bash against the sides and bottom of this river. Over a long time, the moving water and rocks carved this deep **gorge.** The water changed the shape of the land.

water direction

sediment

Water carries sediment. The sediment rubs against rocks in the stream. Over time this makes the larger rocks smooth.

NEXT GENERATION SCIENCE STANDARDS | DISCIPLINARY CORE IDEAS

ESS1.C: The History of Planet Earth
Some events happen very quickly; others occur very slowly, over a time period much longer than one can observe. (2-ESS1-1)

ESS2.A: Earth Materials and Systems
Wind and water can change the shape of the land. (2-ESS2-1)

A **gully** is like a large ditch or small valley. It forms when water erodes soil. Moving water erodes soil more quickly than it can erode rock.

The Narrows in Zion National Park is a gorge.

Wrap It Up!

1. How is the formation of a gorge similar to a gully? How are they different?
2. Why are the rocks in a gorge's river bed likely to have smooth edges?

My science notebook

Wind and Water Move Sand

Grains of sand are tiny pieces of rock. Wind pushes grains of sand up one side of a dune. The sand grains roll down the other side. In time, all of the grains that make up the dune have moved. Wind moves a sand dune much more quickly than it can carve rock.

Water changes a beach much more quickly than it can carve a gorge. Heavy waves during a storm can wash away much of the sand on a beach in a few hours.

NEXT GENERATION SCIENCE STANDARDS | DISCIPLINARY CORE IDEAS

ESS1.C: The History of Planet Earth
Some events happen very quickly; others occur very slowly, over a time period much longer than one can observe. (2-ESS1-1)

ESS2.A: Earth Materials and Systems
Wind and water can change the shape of the land. (2-ESS2-1)

This sand dune is in the Sahara Desert.

Wrap It Up!

1. Explain how these pictures show examples of erosion.
2. Describe how wind and water might affect a sand castle that you build on a beach.

My science notebook

Investigate

Erosion

How can you prevent erosion?

Water can slowly weather and erode rock. It can also quickly erode softer soil and beaches. People try to prevent water from eroding land. In this investigation you will use materials to slow down or prevent erosion.

Materials

NEXT GENERATION SCIENCE STANDARDS | DISCIPLINARY CORE IDEA

ESS1.C: The History of Planet Earth
Some events happen very quickly; others occur very slowly, over a time period much longer than one can observe. (2-ESS1-1)

ESS2.A: Earth Materials and Systems
Wind and water can change the shape of the land. (2-ESS2-1)

1 Make a hill of soil in one end of each tray.

2 Use the measuring cup to pour 100 mL of water over the soil in one tray. Observe and record how the moving water erodes the soil.

3 Use the materials provided to make model plants, fences, and roots and place them on the hill of soil in the second tray.

4 Use the measuring cup to pour 100 mL of water over the soil on the second tray. Observe and record how the moving water erodes the soil.

Wrap It Up!

1. How did pouring the water affect the first hill of soil? How did pouring the water affect the second hill of soil?
2. Compare how the water affected each hill. What was the same? What was different?

My science notebook

Think Like a Scientist

Make Observations

In the investigation on pages 88–89, you saw how water can change the land quickly. Other changes may happen slowly, over a long time. Use the internet or other sources to find evidence of fast and slow changes on Earth's surface. Then, look at each picture, and decide whether it shows a quick change or a slow change.

Wind is blowing the sand and changing the shape of the dune.

NEXT GENERATION SCIENCE STANDARDS | PERFORMANCE EXPECTATION
2-ESS1-1. Use information from several sources to provide evidence that Earth events can occur quickly or slowly.

Melted rock and ash erupt from this volcano.

Rock has weathered and the pieces eroded, leaving this tower shape.

An earthquake has made this building collapse.

A gully has formed in a farmer's field.

Wrap It Up!

1. Tell which changes shown can be fast and which is usually slow.
2. Tell what evidence you used to decide if the change could be fast or slow.

My science notebook

Think Like an Engineer
Case Study

Protecting New Orleans

Problem

You know that water can change land. It can also damage buildings and roads.

The city of New Orleans in Louisiana is built on land that is mostly below sea level. Water in nearby lakes and rivers is higher than the level of the city streets. If the water overflows the banks of those lakes and rivers, it floods streets and buildings.

NEXT GENERATION SCIENCE STANDARDS | DISCIPLINARY CORE IDEAS
ESS1.C: The History of Planet Earth
Some events happen very quickly; others occur very slowly, over a time period much longer than one can observe. (2-ESS1-1)

ESS2.A: Earth Materials and Systems
Wind and water can change the shape of the land. (2-ESS2-1)

Hurricane Katrina was a huge storm. It caused terrible flooding. The flooding destroyed much of the city of New Orleans.

Solution

A **levee** is a barrier that keeps water from flooding into places where it is not wanted. New Orleans is surrounded by levees and flood walls. They are made of earth, concrete, and steel. The levees are very strong. The tall walls keep water out of the city even when storms cause rivers and lakes to overflow.

City of New Orleans

Levee

Levee

Lake Pontchartrain sits on one side of New Orleans. A levee keeps the water in the lake and out of the city.

The Mississippi river flows on the other side of the city. A levee keeps the water in the river and out of the city.

Wrap It Up!

1. How can water cause fast changes that affect a city?
2. Describe how a levee can help control flooding.

My science notebook

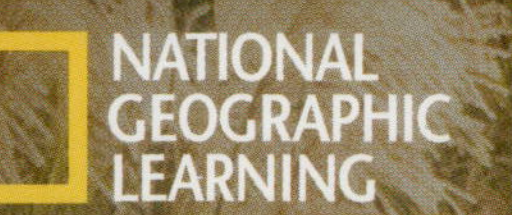

Think Like an Engineer

Compare Solutions

Stop That Runoff!

People have built buildings, roads, and parking lots. They cover a lot of land. Land that is covered with buildings or pavement cannot soak up rainwater. The water has to drain away.

Water running along the ground can cause erosion. It moves soil and rocks from one place to another. It can also wash pollution into ponds and streams.

One way to reduce rainwater **runoff** is to plant a rain garden. A rain garden collects rainwater. It gives water a chance to soak into the ground.

Water runs to the lowest spot in the rain garden. Then it soaks into the ground.

NEXT GENERATION SCIENCE STANDARDS | DISCIPLINARY CORE IDEAS
ETS1.C: Optimizing the Design Solution
Because there is always more than one possible solution to a problem, it is useful to compare and test designs. (secondary to 2-ESS2-1)

NEXT GENERATION SCIENCE STANDARDS | PERFORMANCE EXPECTATION
2-ESS2-1. Compare multiple solutions designed to slow or prevent wind or water from changing the shape of the land.

The rain garden plants can grow naturally without chemicals.

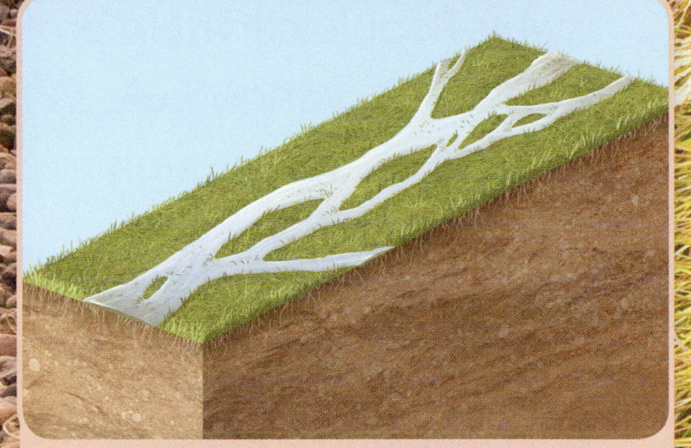

Before the rain garden, water collected on the surface. The water had to run downhill to drain away.

The rain garden is at a low spot so rainwater collects in it.

Stop That Sand

It is always windy at the beach! Wind causes erosion of sand dunes along the shore. Roots of sea grasses hold some of the sand in place so it doesn't blow away. People can help protect sand dunes by building windbreaks. A fence can block the wind enough to keep the sand in place.

Stop That Soil!

Wind erosion can be a problem for farmers, too. The wind can carry away the soil needed for growing crops. Farmers can use fences for windbreaks. They can also plant rows of tall, thick shrubs or trees to block the wind.

Bushes block wind across this field of lavender.

NEXT GENERATION SCIENCE STANDARDS | PERFORMANCE EXPECTATION
2-ESS2-1. Compare multiple solutions designed to slow or prevent wind or water from changing the shape of the land.

Rows of fence keep sand along the shore from blowing away. If you look closely at the picture, you can see that sand dunes have built up all the way to the top of the fence in some areas.

Wrap It Up!

1. Suppose thick grasses grow along a sand dune fence. Explain why that could allow people to remove the fence.
2. Do you think a row of trees or a fence would make a better windbreak for a farmer? Explain your answer.
3. How are rain gardens and windbreaks alike and different?

My science notebook

Understanding Maps

Maps show where things are located. A physical map shows the shapes of land and water. It also shows the kinds of land and water in an area.

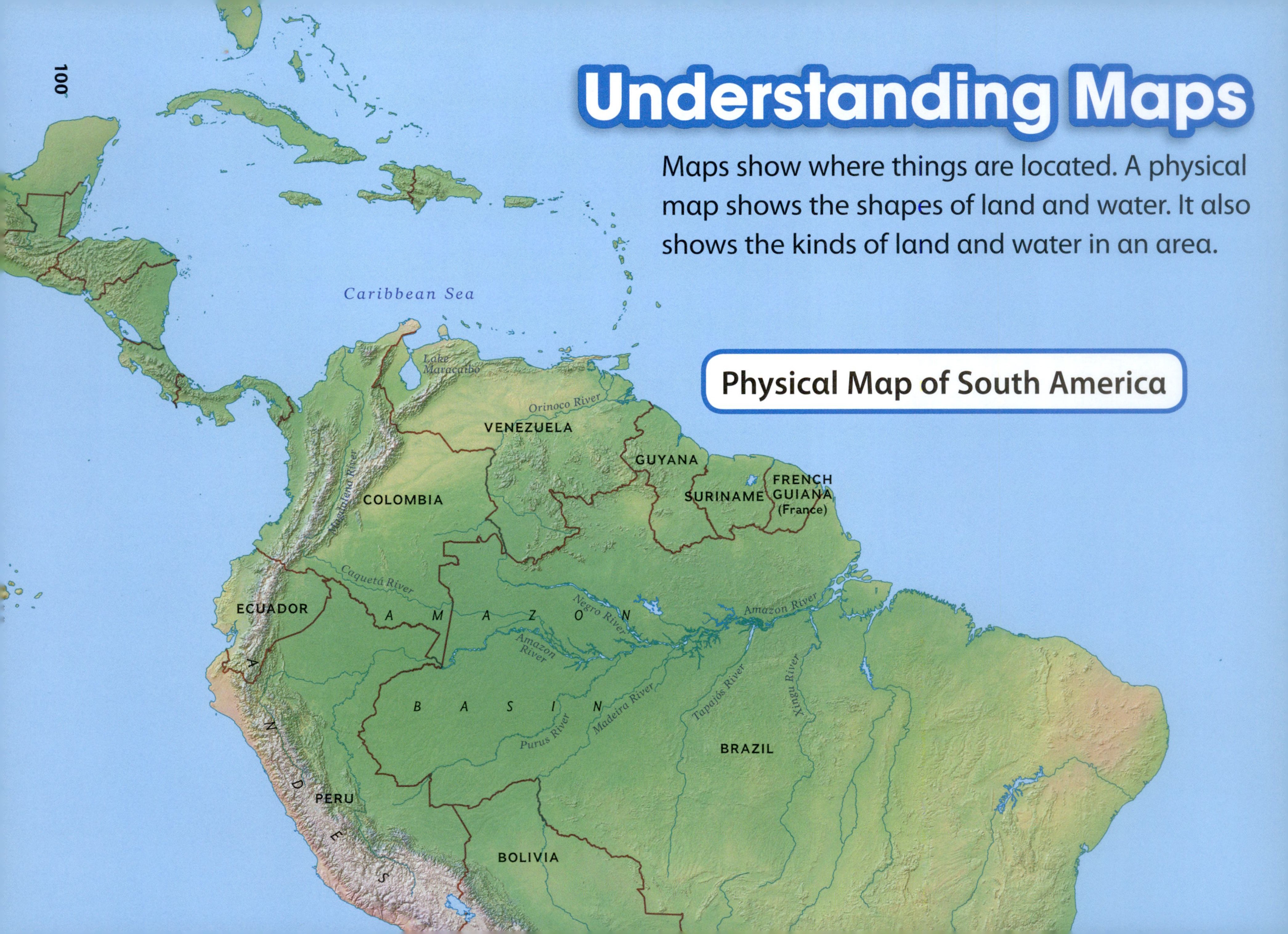

Wrap It Up!

1. Does South America have more forests or grasslands?
2. Where are most of the mountains in South America located?
3. What can you observe about the Amazon River from the map?

My science notebook

NEXT GENERATION SCIENCE STANDARDS | DISCIPLINARY CORE IDEAS
ESS2.B: Plate Tectonics and Large-Scale System Interactions
Maps show where things are located. One can map the shapes and kinds of land and water in any area. (2-ESS2-2)

Rivers and Oceans

Some water on land flows downhill. It collects into small streams. The streams become larger and form **rivers.** Rivers get bigger and move more slowly as they reach flatter ground. Many rivers empty into the **oceans.**

Streams are smaller than rivers. Many of these streams flow quickly like this one.

Some of the water flowing in rivers eventually reaches the ocean. Oceans surround the continents. Most of Earth's surface is covered by oceans.

NEXT GENERATION SCIENCE STANDARDS | DISCIPLINARY CORE IDEAS
ESS2.C: The Roles of Water in Earth's Surface Processes
Water is found in the ocean, rivers, lakes, and ponds. Water exists as solid ice and in liquid form. (2-ESS2-3)

The Amazon River in South America is huge. It carries more water than any other river in the world.

Wrap It Up!

1. Compare streams and rivers. How are they alike? How are they different?
2. Is there more water in Earth's rivers or oceans? Explain.

My science notebook

Lakes and Ponds

Not all surface water on land flows in rivers and streams. Water also collects in low places on land. Low places can fill up like a bowl. Water collects there to form lakes and ponds. A **lake** is a large body of water surrounded by land. A **pond** is a small lake.

Lake Pehoe in Chile

Ponds can dry up if too little rain falls to keep them filled with water.

NEXT GENERATION SCIENCE STANDARDS | DISCIPLINARY CORE IDEAS
ESS2.C: The Roles of Water in Earth's Surface Processes
Water is found in the ocean, rivers, lakes, and ponds. Water exists as solid ice and in liquid form. (2-ESS2-3)

Lakes are different from oceans. Lakes are surrounded by land.

Wrap It Up!

1. How does a lake form?
2. Why is the base of a mountain a likely place for a lake to form?

My science notebook

Make a Model

Our planet has many different kinds of land and water features. Look closely at the photograph on this page. What kinds of land and water do you see?

Design a model.

Scientists use models to show objects and ideas that are hard to see. Draw a model of the land and water features you see in the photographs. One photo shows what this beach in Cornwall, England, looks like from above. The other shows what the area looks like from the ground.

NEXT GENERATION SCIENCE STANDARDS | PERFORMANCE EXPECTATION
2-ESS2-2. Develop a model to represent the shapes and kinds of land and bodies of water in an area.

Build your model.

Use your drawing to help you build your model.

1. Use clay or other materials to make a model of each land and water feature.
2. Build your model on a piece of cardboard.
3. Study your model. Record your observations in your science notebook.

Share your model.

How does your model help you understand how the land is shaped? What other science questions can your model help you answer?

Ice on Earth

Not all Earth's water is liquid. Water can be solid, too. Solid water is called ice. Some places on Earth stay too warm for ice to form. Other places such as the North Pole and the South Pole are cold. Water is frozen year round at the Poles.

Mountains are usually cold places. There snow can build up into layers. Over time the icy layers become very thick. These thick sheets of ice can slowly move downhill like rivers of ice. These moving ice sheets are called **glaciers.** Sometimes glaciers reach the ocean. When this happens, large chunks of ice called **icebergs** can break off and float away.

Large chunks of ice floating in the ocean are called icebergs.

NEXT GENERATION SCIENCE STANDARDS | DISCIPLINARY CORE IDEAS
ESS2.C: The Roles of Water in Earth's Surface Processes Processes
Water is found in the ocean, rivers, lakes, and ponds. Water exists as solid ice and in liquid form. (2-ESS2-3)

A glacier forms from packed snow that hardens to solid ice. Like rivers, glaciers move downhill.

Wrap It Up!

1. Where can Earth's water be found?
2. Where can Earth's water be found as ice?

My science notebook

Obtain Information

The world map shows land and water. Use the map to find rivers, lakes, and oceans. Use the map to find water that is frozen as ice.

NEXT GENERATION SCIENCE STANDARDS | PERFORMANCE EXPECTATION
2-ESS2-3. Obtain information to identify where water is found on Earth and that it can be solid or liquid.

Land and Water on Earth

Wrap It Up!

1. Where do you see liquid water in Africa?
2. In which places do you see the most liquid water on Earth?
3. What places on Earth have the most ice?

My science notebook

Glaciologist

Erin Pettit is a glaciologist. She studies how ice sheets have grown and shrunk. Glaciers change over thousands of years. They give clues to how Earth's climate has changed.

Glaciers can help scientists predict changes in many other places. When glaciers shrink, ocean levels are higher. Higher ocean levels affect islands and coasts. Erin Pettit studies the places where glaciers melt and break off into the ocean. She collects data. She notices if the changes are happening faster.

NEXT GENERATION SCIENCE STANDARDS | CONNECTIONS TO NATURE OF SCIENCE
Scientific Knowledge Is Based on Empirical Evidence
Scientists look for patterns and order when making observations about the world.

Explorer

Erin Pettit studies glaciers. She also teaches young people about the science she works on. She started a wilderness science program for high school girls called Girls on Ice.

To study glaciers, Erin spends a lot of time on the ice. Her work requires special clothing and equipment.

Glossary

A

absorb (ab-ZORB)

To absorb is to soak up liquid. (p. 26)

C

coast (KŌST)

A coast is land that is next to an ocean. (p. 64)

D

depend (dē-PEND)

To depend is to need something in order to live. (p. 44)

E

earthquake (URTH-kwāk)

An earthquake is a sudden shaking of the ground caused by land moving. (p. 76)

erosion (i-RŌ-zhun)

Erosion is the movement of rocks or soil caused by wind, water, or ice. (p. 80)

erupt (ē-RUPT)

Materials from inside Earth erupt from a volcano. (p. 78)

The ocean meets the land at this rocky **coast.**

F

flexible (FLEX-uh-bul)
To be flexible is to be able to bend without breaking. (p. 20)

G

glacier (GLĀ-shur)
A glacier is a large, moving sheet of ice. (p. 108)

gorge (GORJ)
A gorge is a tall, narrow passage made of rock. (p. 84)

grassland (GRAS-land)
A grassland is a dry place where many grasses but few trees grow. (p. 68)

gully (GUH-lē)
A gully is a crack formed in the Earth by running water. (p. 86)

I

iceberg (ĪS-burg)
An iceberg is a large chunk of floating ice. (p. 108)

L

lake (LĀK)
A lake is a large body of water surrounded by land. (p. 104)

levee (LE-vē)
A levee is a wall built to protect cities from flooding. (p. 94)

liquid (LIK-wid)
A liquid is matter that takes the shape of its container. (p. 6)

The climber's rope is **flexible.**

M

matter (MA-tur)
Matter is anything that takes up space. (p. 4)

O

ocean (Ō-shun)
A large body of water not surrounded by land. (p. 102)

P

pollen (PAH-lin)
Pollen is made and exchanged by flowering plants to make new plants. (p. 50)

pollinate (PAH-li-nāt)
To pollinate is to move pollen from a part of a flower that makes pollen to a part of a flower that makes seeds. (p. 50)

pond (PAHND)
A pond is a small body of still water. (p. 104)

property (PRAH-per-tē)
A property is something about an object that you can observe. (p. 12)

R

river (RI-vur)
A river is a large natural stream of water. (p. 102)

runoff (RUN-awf)
Water that flows over the ground after a rain. (p. 96)

Bees and other animals **pollinate** flowers.

S

sediment (SED-uh-ment)
Sediment is tiny pieces of rock. (p. 82)

solid (SAH-lid)
A solid is matter that has its own shape. (p. 8)

T

texture (TEKS-chur)
Texture is the way an object feels. (p. 16)

V

volcano (vol-KĀ-nō)
A volcano is an opening on Earth from which lava flows. (p. 78)

W

weathering (WE-thur-ing)
Weathering is the breaking of rocks into smaller pieces. (p. 80)

wetland (WET-land)
A wetland is land that is covered with water some of the time. (p. 66)

Glowing hot lava erupts from a **volcano.**

Index

D

E

F

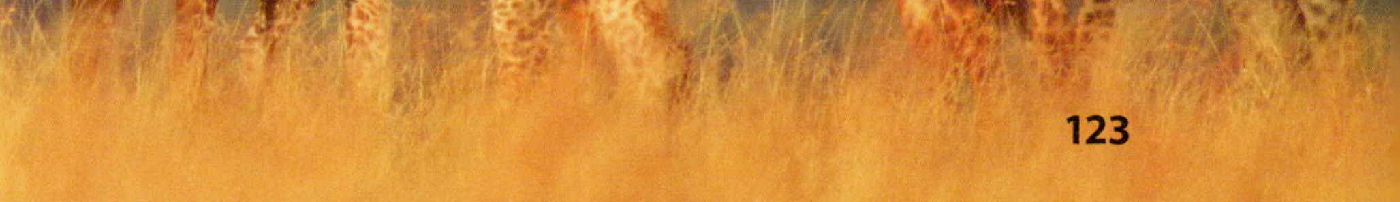

R

S

T

V

W

Y

Z

Photographic and Illustrator Credits

Front Matter

Title Page ©John Eastcott and Yva Momatiuk/National Geographic Creative. **ii–iii** ©Brook Tyler Photography/Flickr/Getty Images. **iv–v** ©Ben Cranke/The Image Bank/Getty Images. **v–vi** ©Robin Wilson Photography/Flickr Open/Getty Images.

Physical Science: Structure and Properties of Matter

2–3 ©Piriya Photography/Flickr Select/Getty Images. **4–5** ©Alice Artime/Flickr Open/Gety Images. **6–7** ©Robbie George/National Geographic Creative. **7** (tr) ©Volkmar K. Wentzel/National Geographic Creative. **8–9** ©Spring Images/Alamy. **9** (tr) ©Ryan/Beyer/The Image Bank/Getty Images. **10** (cl) ©National Geographic Learning. (c) ©Jeanine Childs/National Geographic Learning. (cr) ©National Geographic School Publishing. **10–11** ©KentarooTryman/Folio Images/Getty Images. **11** (tr)©Jeanine Childs/National Geographic Learning. (cr) ©Jeanine Childs/National Geographic Learning. **12–13** ©Danita Delimont/Gallo Images/Getty Images. **14–15** ©Douglas Peebles Photography/Alamy. **16–17** ©Doug Marshall/Getty Images. **17** (tr) ©Jeanine Childs/National Geographic Learning. **18–19** ©Kate Thompson/National Geographic Creative. **19** (tr) ©Jeanine Childs/National Geographic Learning. **20–21** ©Greg Epperson/Getty Images. **21** (tl) ©William Alberet Allard/National Geographic Creative. (tr) ©iStockphoto.com/SweetyMommy. **22** (cl) ©David Madison/Photodisc/Getty Images. **22–23** ©Sami Suni/Vetta/Getty Images. **24–25** ©Linda Burgess/Photolibrary/Getty Images. **26** (cl) ©National Geographic Learning. (c) ©National Geographic School Publishing. (c) ©Jeanine Childs/National Geographic Learning. (cr) ©National Geographic Learning. (b) ©Jeanine Childs/National Geographic Learning. **26–27** ©Adam Gault/OJO Images/Getty Images. **27** (tl) ©Jeanine Childs/National Geographic Learning. (cl) ©Jeanine Childs/National Geographic Learning. **28–29** ©Lonely Planet Images/Getty Images. **29** (tr) ©Duncan Selby/Alamy. **30–31** ©Jeffrey Coolidge/Photodisc/Getty Images. **32–33** ©Kevin Schafer/Minden Pictures. **33** (tr) ©National Geographic School Publishing. **34–35** ©Tetra Images/Corbis. **35** ©National Geographic School Publishing. **36–37** ©Marjorie McBride/Alamy. **37** (tl) ©Image Source/Getty Images. **37** (tr) ©Eye Ubiquitous/Alamy. **38–39** ©Alexey Tkachenko/E+/Getty Images. **39** (r) ©Lauri Patterson/E+/Getty Images. (l) ©Micheal S. Quinton/National Geographic Creative. **40–41** ©James Duncan Davidson. **41** (tr) ©National Geographic Creative.

Life Science: Interdependent Relationships in Ecosystems

42–43 ©Joel Sartore/National Geographic Creative. **44–45** ©Raul Touzon/National Geographic Creative. **45** (tr) ©Ernie Janes/Nature Picture Library. **46** (cl) ©National Geographic Learning. (cr) ©National Geographic Learning. (bl) ©National Geographic Learning. (br) ©National Geographic Learning. **46–47** ©Joe Petersburger/National Geographic Creative. **47** (tr) ©National Geographic Learning. (tl) ©National Geographic Learning. **48–49** ©Jacky Parker/Alamy. **50** (bl) ©Geopge Grall/National Geographic Creative. **50–51** ©Nature Production/Nature Picture Library. **51** (tr) ©Genevieve Vallee/Alamy. **52–53** ©Dino Martins. **53** (tr) ©Dino Martins. (tr) ©Dino Martins. (cr) ©Dino Martins. (br) ©Dino Martins. **54** (bc) ©Voorspoels Kurt/Arterra Picture Library/Alamy. **54–55** ©Cecilia Lewis/Courtesy of Dino Martins. **55** (tr) ©Dino Martins. **56** (bl) ©Elliot Neep/Oxford Scientific/Elliot Neep/Getty Images. © Torbjörn Arvidson/Matton Collection/Corbis. **56–57** ©P. Schuetz/Blickwinkel/AGE Fotostock. **58** (c) ©Andrew Darrington/Alamy. **58–59** ©Andrey Nekrasov/AGE Fotostock. **60** (tl) ©Ken Kinzie/National Geographic Learning. (tr) ©Jianghaistudio/Shutterstock.com. (cl) ©Ken Kinzie/National Geographic Learning. (cr) ©Siede Preis/Photodisc/Getty Images. **60–61** ©Ruud Morijn Photographer/Shutterstock.com. **62–63** ©Brian J.Skerry/National Geographic Creative. **63** (tr) ©Doug Cheeseman/Oxford Scientific/Getty Images. (cr) ©blickwinkel/McPhoto/MO/Alamy. **64** (bl) ©NSP–RF/Alamy. (br) ©Jesse Cancelmo/Alamy. **64–65** ©Tony Campbell/Shutterstock.com. **66–67** ©Tim Graham/Alamy. **68–69** ©Sailer/blickwinkel/Alamy. **69** (tr) ©Dave Watts/Alamy. (cr) ©Martin Ruegner/The Image Bank/Getty Images. **70** (tr) ©Anup Shah/Stone/Getty Images. (tr) ©Christopher Herwig/Lonely Planet Images/Getty Images. (cr) ©Visual&Written SL/Alamy. (cr) ©Heinrich van den Berg/Gallo Images/Getty Images. (br) ©Frans Lanting/National Geographic Creative. (br) ©age fotostock/Alamy. **71** (tc) ©LOOK Die Bildagentur der Fotografen GmbH/Alamy. (tr) ©Joachim Hiltmann/imag/AGE Fotostock. (cl) ©Photoshot Holdings Ltd/Alamy. (bl) ©Image Source/Corbis. (bc) ©Graham Robertson/Auscape/Minden Pictures. (br) ©Reinhard Dirscherl/Visuals Unlimited/Corbis. **72–73** ©Tim Laman/National Geographic Creative. **73** (tr) ©Tim Laman/National Geographic Creative. (cr) ©Tim Laman/National Geographic Creative. (cr) ©Tim Laman/National Geographic Creative. (br) ©Tim Laman/National Geographic Creative.

Earth Science: Earth's Systems: Processes that Shape the Earth

74–75 ©Momatiuk–Eastcott/Corbis. **76** (bc) ©Tony Wheeler/Lonely Planet Images/Getty Images. **76–77** ©Anthony Asael/Art in All of Us/Corbis. **78** (cl) ©Gary Rosenquist. (cr) ©Science Source/USGS/Photo Researchers/Getty Images. **78–79** ©Sunset Avenue Productions/Digital Vision/Getty Images. **80–81** ©Chris Hepburn/The Image Bank/Getty Images. **81** (cr) ©National Geographic Learning. **82–83**

©Melissa Farlow/National Geographic Creative. **83** (tr) ©Robert Cable/ Photodisc/Getty Images. **84–85** ©Greg Winston/National Geographic Creative. **85** (tr) ©De Meester Johan/Arterra Picture Library/Alamy. **86** (c) ©Cheryl Casey/Shutterstock.com. **86–87** ©Sergio Pitamitz/ Corbis. **88** (tc) ©Michael Goss Photography/National Geographic Learning. (tr) ©iStockphoto.com/Ljupco. (cl) ©National Geographic School Publishing. (c) ©National Geographic School Publishing. (cr) ©Michael Goss Photography/National Geographic Learning. (bl) ©Michael Goss Photography/National Geographic Learning. (br) ©National Geogrpahic Learning. (tl) ©National Geographic School Publishing. **88–89** ©Nigel Cattlin/Visuals Unlimited/Corbis. **89** (tl) ©Michael Goss Photography/National Geographic Learning. (cl) ©Michael Goss Photography/National Geographic Learning. **90** (c) ©Thomas Northcut/The Image Bank/Getty Images. **90–91** ©Susan/ Neil Silverman/FogStock/Alamy. **91** (tl) ©Ross Armstrong/Alamy. (tr) ©Fotosearch/Getty Images. (cl) ©Garry Gay/Photographer's Choice/ Getty Images. (cr) ©De Meester Johan/Arterra Picture Library/Alamy. **92–93** ©Everett Collection/Everett Collection Inc./Age fotostock. **94–95** ©AP Images/Dave Martin. **96–97** ©Saxon Holt/Alamy. **98** (cr) ©Camille Moirenc/Hemis/Corbis. **98–99** ©PlainPicture/Glow Images. **102** (bl) Bates Littlehales/National Geographic Creative. (br) ©Richard Du Toit/Minden Pictures. **102–103** ©Layne Kennedy/Corbis. **104** (cl) ©JLR Photography/Shutterstock.com. **104–105** ©Fotofeeling/ Westend61 GmbH/Alamy. 106 (bc) ©Resnyc/Flickr/Getty Images. **106–107** ©Peter Barritt/SuperStock/Alamy. **107** (bc) ©Andy Stothert/ Britain On View/Getty Images. **108** (tr) ©nagelestock.com/Alamy. **108–109** ©James Davis Photography/Alamy. **110–111** ©Mint Images/ Paul Edmondson/Vetta/Getty Images. **112–113** ©Adam Taylor, Courtesy of Erin Pettit. **113** (tr) ©Adam Taylor, Courtesy of Erin Pettit.

End Matter

114–115 ©Rich Reid/National Geographic Creative. **116–117** ©Tyler Stableford/Digital Vision/Getty Images. **118–119** ©Konrad Wothe/ Minden Pictures. **120–121** ©Toshi Sasaki/Stone/Getty Images. **122–123** ©WLDavies/Vetta/Getty Images. **124–125** ©Tongho58/Flickr/Getty Images. **126–127** ©subtik/E+/Getty Images.

Illustration Credits

Unless otherwise indicated, all illustrations were created by Precision Graphics and all maps were created by Mapping Specialists.

Content Consultants

Randy L. Bell, Ph.D.
Associate Dean and Professor of Science Education, College of Education, Oregon State University

Malcolm B. Butler, Ph.D.
Associate Professor of Science Education, School of Teaching, Learning and Leadership, University of Central Florida

Kathy Cabe Trundle, Ph.D.
Department Head and Professor, STEM Education, North Carolina State University

Judith S. Lederman, Ph.D.
Associate Professor and Director of Teacher Education, Illinois Institute of Technology

Acknowledgments
Grateful acknowledgment is given to the authors, artists, photographers, museums, publishers, and agents for permission to reprint copyrighted material. Every effort has been made to secure the appropriate permission. If any omissions have been made or if corrections are required, please contact the Publisher.

Photographic and Illustrator Credits
Front cover ©John Eastcott and Yva Momatiuk/National Geographic Creative. **Back cover** ©Adam Taylor, Courtesy of Erin Pettit. (tr) ©Adam Taylor, Courtesy of Erin Pettit.

Acknowledgments and credits continued on page 128.

Visit National Geographic Learning online at NGL.Cengage.com

Visit our corporate website at www.cengage.com

Printed in the USA.
RR Donnelley

ISBN: 978-12858-46347

16 17 18 19 20 21 22 23

10 9 8 7